Library of Congress Cataloging-in-Publication Data

Watts, Barrie.
    Ladybug.

     (Stopwatch books)
     Includes index.
     Summary: Photographs, drawings, and text on
two different levels of difficulty follow the
stages of development of the ladybug, from
mating and hatching of the larvae to the growth of
the larvae into adult ladybugs.
     1. Ladybirds — Juvenile literature. [1. Ladybugs]
I. Title. II. Series.
QL596.C65W38   1987   595.76'9   86-31476
ISBN 0-382-09441-7
ISBN 0-382-09437-9 (lib. bdg.)

First published by A & C Black (Publishers) Limited
35 Bedford Row, London WC1R 4JH

© 1987 Barrie Watts

Published in the United States in 1987
by Silver Burdett Press
Morristown, New Jersey

Acknowledgements
The artwork is by Helen Senior
The publishers would like to thank Jean Imrie and Simon Ryder for their help and advice.

Ladybug

Barrie Watts

Stopwatch books

 Silver Burdett Press • Morristown, New Jersey

Here is a ladybug.

Have you ever seen a ladybug? They live in parks and gardens. Ladybugs are a sort of beetle. There are lots of different kinds of ladybugs. This one has seven spots.

This ladybug is eating an aphid. Aphids are a ladybug's favorite food.

This book will tell you about the life of a ladybug.

The ladybug finds a mate.

In spring the ladybug is ready to mate. He uses his antennae to find a female ladybug. He knows when he has found a female because of her special smell.

Look at the photograph. These ladybugs are mating. The male is behind the female.

A week later, the female lays her eggs.

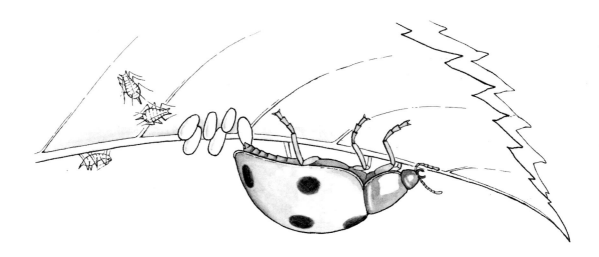

She always lays the eggs near some aphids so that when the larvae hatch, they will have some food to eat.
After the female ladybug has laid the eggs, she will die.

The larvae hatch out of the eggs.

The ladybug's eggs are bright yellow. They are glued firmly to a leaf. After a few days, they turn white. This means that they are ready to hatch.

These tiny creatures are the larvae of the ladybug. They have just hatched.

When the larvae come out of the eggs they are white, but they soon turn black. Each larva is about as big as the top of a pin.

The larvae grow fast.

As soon as the larvae have hatched, they start looking for food. These larvae are one day old.

As a larva gets bigger, its skin gets too tight. Then its skin starts to split, like this.

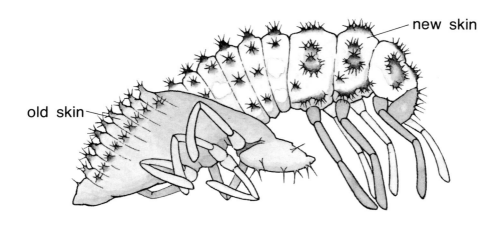

new skin

old skin

There is a new skin underneath. Soon the larva's skin will split again and orange markings will appear on its body. The ladybug larva will change its skin four times before it is fully grown.

The larva is always hungry.

This drawing shows a larva from underneath.

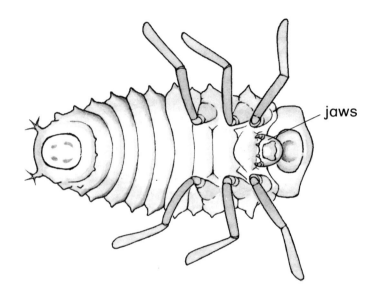

jaws

The larva's body is divided into segments. It has six legs, and a tiny sucker at the end of its body.

Look at the big photograph. This larva is looking for aphids. It is always hungry. It eats about 30 aphids every day. The larva uses its strong jaws to grip the aphid and suck the juice out of its body.

The larva changes into a pupa.

The larva has grown very fat. Now it has stopped eating and it will not grow any bigger. It is ready to change into a pupa. The larva finds a safe place to rest under a leaf. It fixes itself to the stem with sticky glue from its rear end.

After two hours, the skin of the larva starts to split down the back. Look at the drawing.

The pupa inside is beginning to show through.

The pupa has a dry, hard shell.

When the skin has split, the pupa looks like this.

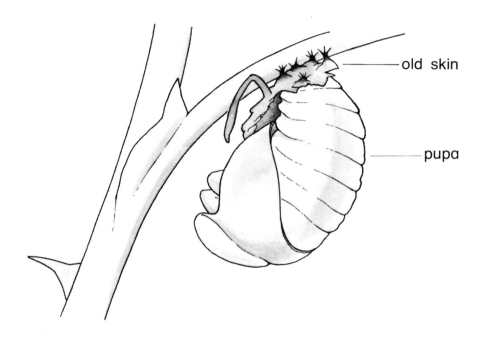

old skin

pupa

It is soft and orange. The old skin is still attached
to the end of the pupa.

Look at the photograph. After one day, the pupa has
a dry hard shell. There are black markings on it.
The pupa stays very still. Inside the shell,
the pupa is changing into an adult ladybug.

The ladybug comes out of the pupa.

Five days later, the pupa has finished changing into a ladybug. It is ready to come out of the pupal shell. The ladybug pushes from the inside until the shell splits open. Then it struggles out, head first.

Look at the big photograph. The ladybug is almost out. It takes about five minutes to struggle from the shell.

The ladybug's body is pale orange, and soft and damp. It needs to crawl to a safe place until its body is dry.

The ladybug dries its wings.

The ladybug is waiting for its body to dry. Slowly it opens its wings so that they can dry, too. The ladybug cannot fly yet.

Look at the big photograph. Can you see the spots appearing on the ladybug's wing cases?

Twelve hours later, the ladybug has all its spots.

Its wing cases are hard and smooth. They turn darker red as the ladybug gets older.

The ladybug flies for the first time.

The ladybug waits for a warm, sunny day. Then it gets ready to fly for the first time. It cannot take off from the ground so it looks for a grass stalk to climb.

This ladybug is getting ready to take-off.

Ladybugs do not fly very fast. They open their red wing cases and beat their wings quickly up and down.

The ladybug rests through the winter.

At the end of the summer, the ladybug looks for a place to spend the winter. It will sleep through the winter under a pile of leaves.

Next spring the ladybug will wake up. Then it will look for a mate.

What do you think will happen then?

Do you remember how the ladybug came from an egg?
See if you can tell the story in your own words.
You can use these pictures to help you.

3

6

Look for ladybugs on a warm summer day. See if